Amy Greenwell
Garden
Ethnobotanical
Guide to

NATIVE HAWAIIAN PLANTS

& Polynesian-Introduced Plants

Amy Greenwell Garden Ethnobotanical Guide to

NATIVE HAWAIIAN PLANTS

& Polynesian-Introduced Plants

Noa Kekuewa Lincoln

With contributions by Peter Van Dyke, Brian Kiyabu,
Clyde Imada, George Staples, & Manuel Rego

Photography by Noa Lincoln, R. Kealohapauʻole Manakū,
Clyde Imada, & Bernice Akamine

BISHOP MUSEUM
PRESS

HONOLULU, HAWAIʻI

This publication was supported by a grant from the Council for Native Hawaiian Advancement and the Hawai'i Tourism Authority, and by a grant from the Nadao & Mieko Yoshinaga Foundation.

Further support came from the Office of Innovation and Improvement of the U.S. Department of Education. Education through Cultural & Historical Organizations, also known as ECHO, provides educational enrichment to Native and non-Native children and lifelong learners. The project was partially funded by the U.S. Department of Education.

This project was further funded under the Native Hawaiian Culture and Arts Program. The views and conclusions contained in this document are those of the authors and should not be interpreted as representing the opinions or policies of the U.S. Government. Mention of trade names or commercial products does not constitute their endorsement by the U.S. Government.

The image of the Hawaiian Islands on page 3 is used courtesy of NASA and can be found on the Visible Earth (http://www.visibleearth.nasa.gov/).

Bishop Museum Press
1525 Bernice Street
Honolulu, Hawai'i 96817
www.bishopmuseum.org/press

ISBN: 978-1-58178-092-5

Design by Nancy Watanabe

Printed in China

Library of Congress Cataloging-in-Publication Data

Amy Greenwell Garden ethnobotanical guide to native Hawaiian plants & Polynesian-introduced plants / compiled by Noa Lincoln ; with contributions by Peter Van Dyke ... [et al.].
    p. cm.
  Includes index.
  ISBN 978-1-58178-092-5 (pbk. : alk. paper)
  1. Plants--Hawaii--Handbooks, manuals, etc. 2. Ethnobotany--Hawaii--Handbooks, manuals, etc.
3. Amy B.H. Greenwell Ethnobotanical Garden--Guidebooks. I. Greenwell, Amy Beatrice Holdsworth.
II. Lincoln, Noa. III. Van Dyke, Peter. IV. Title: Ethnobotanical guide to native Hawaiian plants and Polynesian-introduced plants.
  QK473.H4.A53 2009
  581.9969--dc22

                    2008044382

CONTENTS

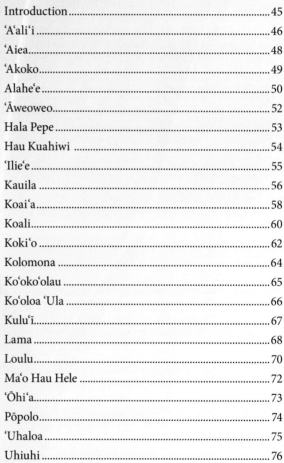

POLYNESIAN-INTRODUCED CROPS

WET FOREST ZONE

AMY B. H. GREENWELL ETHNOBOTANICAL GARDEN

The plants found is this guide, along with over 100 other native species, can all be seen at the Amy B. H. Greenwell Ethnobotanical Garden, in Captain Cook on Hawaiʻi Island, 12 miles south of Kailua-Kona. The Garden is divided into the four zones outlined in this book.

TO BEGIN A SELF-GUIDED TOUR:

We suggest you begin your visit by reading the three-page introduction to better understand the significance of the Garden. When you are ready, follow any path you choose, but pay attention to the garden zones to easily find the plants in this guide.

- Restrooms are in the buildings near the parking lot.
- Drinking water can be found outside the bathroom. Nice and cold.
- Rain is frequent in the afternoons at the Garden. If you need an umbrella, ask a Garden staff person about borrowing one of ours. They're good for the sun, too.
- Mosquitoes are occasionally a problem, especially in the Wet Forest Zone. The Garden does not offer repellent, although we do have some natural remedies for the bites.
- Archaeology is found throughout the Garden so please act respectfully.
- Donations are gladly accepted to support our education and interpretive programs, expand our native collections, and maintain the Garden.

PLEASE FOLLOW OUR THREE RULES:

- DO NOT pick or remove ANY plant material from the Garden.
- DO NOT move any rocks or walk on the rock walls.
- DO stay on the trails and lawn.

The plants in this book can be seen at the Amy B. H. Greenwell Ethnobotanical Garden in Captain Cook on Hawai'i Island. This book is also a useful field guide to many of the native plants seen around the islands. For convenience the plants are separated into four categories: Coastal Zone, Dry Forest Zone, Polynesian-Introduced Crops, and Wet Forest Zone. Within these categories the plants are organized alphabetically by their Hawaiian name.

The ethnobotanical description of each species gives a physical and habitat description as well as traditional and modern uses of the plants.

Plants are known by different names. This box shows the order of names as they appear in the text.

Hawaiian Name (pronunciation)	ULU (OO-LOO)
Common Name	Breadfruit
Scientific Name	*Artocarpus altilis*
Family Name	Moraceae (Fig Family)
Origin	Polynesian-Introduced
Status	(only shows if plant has a rarity status)

INTRODUCTION

FORMATION OF AN ARCHIPELAGO

To appreciate the significance of Hawaiian ethnobotany, it is important to know the origins of both Hawai'i and the Hawaiian people. The main Hawaiian Islands are located at the southeastern end of a chain of islands and seamounts that extends 3,200 miles, almost to the tip of the Aleutian Islands near Alaska. These original Hawaiian Islands started to form 75 to 85 million years ago over a stationary "hotspot" within the earth. This hotspot melted through the earth's crust and caused the volcanic activity responsible for each island. The earth's crust under Hawai'i, known as the Pacific Tectonic Plate, drifted to the northwest by about 3 to 4 inches a year, slowly moving each new island away from the stationary hotspot and creating a chain of islands.

As each island aged, the forces of erosion—ocean waves, winds, and rains—began to take their toll. The oldest Hawaiian Islands have completely eroded away and become underwater mountains, known as the Emperor Seamounts. There are also 132 tiny atolls and shoals, recently designated the Northwestern Hawaiian Islands National Monument, one of the largest environmental preserves in the world. Finally, there are the eight major islands that the world knows as Hawai'i.

Ni'ihau, the oldest of the eight, is about 6 million years old. Hawai'i Island is the youngest of the archipelago, the oldest rocks on its surface being less than half a million years old.

The hotspot is currently centered approximately 30 miles off the southern coast of Hawai'i Island, and has two major vents: Kīlauea, located on the island itself, known as the most active volcano in the world, and Lō'ihi, an underwater volcano off the island's southern coast that will someday become a new island, though not for many thousands of years.

Natural History

Like all volcanic islands, the first islands of the Hawaiian archipelago were born bare and lifeless. To complicate matters, Hawai'i has the distinction of being the most remote land mass in the world, located over 2,000 miles from the nearest high islands to the south, over 2,500 miles from North America, and over 3,000 miles from the major Philippine Islands. Therefore the plant and animal colonizers of these new islands had to cross at least 2,000 miles of the mighty Pacific Ocean and, by chance, find the tiny island. Once a new species arrived, it had to find a suitable place to live, and it had to find a mate— two difficult tasks on a barren island. It is estimated that a new species colonized the archipelago only once every 50,000 years. Due to these obstacles many plants and animals never became established at all. Case in point: Hawai'i has only two native mammals, the Hoary Bat and the Monk Seal, and no native reptiles.

Ferns, whose tiny spores are dispersed worldwide via jet streams, were probably the first colonizers of the Hawaiian Islands. A few other plant species may have arrived by wind, but migrating sea birds transported about 95% of Hawai'i's native flowering plants. Birds carry seeds of consumed fruit in their gut as well as unwanted hitchhikers in their feathers and in mud stuck to their feet. Additional plants may have floated to the islands, but the direction of the currents and the extreme isolation of the islands make this an unlikely event. All in all, scientists estimate that only 700 original plant, animal, and insect species established themselves in Hawai'i. These species that inhabit an area without the aid of humans are known as **indigenous**.

Once a species establishes itself, it begins to spread out across the landscape. In Hawai'i this presented a problem, because as a species moved into a new area, it found itself in a drastically different habitat. Hawai'i houses nearly every habitat known to man, and also boasts unique environments such as Wai'ale'ale on Kaua'i, the wettest place on earth. In the Hawaiian Islands mere miles can separate rain forest and desert conditions. As indigenous colonizers moved into new areas, they adapted to form more than 6,000 unique Hawaiian species that are **endemic**, meaning that they are found nowhere else in the world. Hawai'i has one of the highest rates of endemism in the world. These were periods during the 75 million years of Hawai'i's history when the islands existed above sea level and no Hawaiian plants survived. But whenever land rose above the sea, the process of colonization and adaptation continued with species constantly island-hopping as new islands were formed and old ones sank into the sea.

Polynesian History

The ancestors of the Polynesians came from southeastern Asia. It is agreed that these people started making open sea journeys to the south 5,000 to 6,000 years ago. Having mingled with the locals both

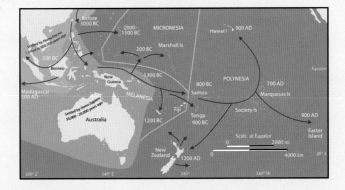

culturally and genetically in the islands of Southeast Asia and Near Oceania, these now Austronesian people left behind a remarkable heritage commonly referred to as the Lapita culture, which thrived 3,500 years ago, primarily in Papua New Guinea.

The Lapita eventually continued their migration east, and by around 1000 B.C. they had colonized Samoa and Tonga. Oddly, in the Samoan area contact with the western islands dwindled, and for nearly 1,000 years the language, mythology, and political structure of the Polynesian culture was developed. Around the time of Christ, the Samoans began making incredibly long sea voyages to the far reaches of the Pacific Ocean. Polynesia is roughly defined by a triangle connecting Hawaiʻi, Aotearoa (New Zealand), and Rapa Nui (Easter Island).

The first people to Hawaiʻi most likely came between 300 and 700 A.D. from the Marquesas, which are the nearest high islands to the south. By this time the Polynesians had perfected their sea voyaging skills, and knew well how to navigate the vast Pacific Ocean, using a combination of astronomy and the reading of weather patterns and ocean birds. They also knew what plants and animals were needed to colonize a new island and the best way to transport those resources. The Polynesians brought to Hawaiʻi about 27 different plant species and 4 animals: dogs, chickens, pigs, and, accidentally, rats. Polynesian culture flourished in Hawaiʻi because of the plants and animals the voyaging Polynesians brought with them in their canoes and because they learned how to use the new plants that they found here.

COASTAL ZONE

THE COASTAL ZONE is roughly defined as any habitat below 500 feet elevation. Within this range we find more specific habitats, such as strands (sites affected by salt spray), coastal plains (sites under 500 feet unaffected by salt spray and seawater), and coastal wetlands (brackish water marshes affected by springs and tides).

The coastal habitats can vary greatly from island to island. On the younger islands the often low rainfall near the shorelines has done little to erode the lava rock, and only sparse plants grow in the cracks formed during lava flows. The coast itself is still rough and rocky. In these young areas fresh and brackish springs are often an oasis for wild plants. These springs exist near the coast on the larger Hawaiian Islands. The large mountains act as aquifers, storing water within their porous volcanic rock and slowly releasing it primarily as coastal or underwater springs. These coastal springs were essential to early Hawaiian life as sources of water for drinking and irrigation. The springs also create wetland habitats that support a variety of birds, such as stilts and plovers, as well as wetland plants, such as sedges and succulents.

On mid-aged islands (1 to 3 million years old) that have eroded much more, streams bring soil deposits from the wetter uplands, and sand and reef begin to protect the coastline. These factors produce a number of diverse habitats that support a broad spectrum of species. New habitats arise such as lagoons, salt flats, and inland marshes. The oldest islands continue this trend until there is ample soil at the coastline and offshore reefs absorb much of the ocean's surf. These factors allow some upland plants to encroach the shoreline, effectively eliminating much of the coastal zone. These oldest islands also lose much of their coastal habitat to geological aging, creating for example the steep sea cliffs on their windward sides.

The coastal plants of Hawai'i are hardy individuals that withstand the drought conditions and high salt content of the coastal plains. Common features of these plants are drought and salt tolerance, bright silvery leaves that reflect sunlight, and small hairs on the plant that help to capture moisture and shield the sun. Coastal plants are often slow-growing and low-to-the-ground due to the harsh conditions. In many cases Hawaiian plants will grow as a crawling ground cover at the coast but as an upright tree in other habitats.

'AE'AE (ay-ay)
Bacopa monnieri
Scrophulariaceae (Figwort Family)
Indigenous

This crawler forms mats of vegetation around marshes and mud flats and along the edges of streams near the coast. *'Ae'ae* makes a fine ground cover in areas where water is abundant. Hawaiian ground covers fill an important niche in the habitat, especially in coastal and wetland areas. They provide nesting grounds for birds, reduce erosion, facilitate silt settlement, and, in some cases, are the only real vegetation present. These attractive and hardy plants are increasingly being used for private landscaping.

'AHU'AWA (ah-who-ah-vah)
Cyperus javanicus
Cyperaceae (Sedge Family)
Indigenous

This sedge is commonly found in a variety
of wetland habitats, from the coast to the
uplands. *'Ahu'awa* tolerates salty water
and soils. The sharp, serrated edges of the
leaves make for rough handling, but that
didn't stop the Hawaiians from weaving
'ahu'awa leaves into a fine, pliable mesh.
The long flower stalks were also pounded
into fiber and braided for cordage.

'ĀKIA (ah-key-ah)
Wikstroemia uva-ursi
Thymelaeaceae ('Ākia Family)
Endemic

'Ākia is relatively rare in the coastal and dry forest habitats, but this small, attractive shrub is becoming a popular ornamental for native landscapers. In addition to being a fish toxin, certain varieties of *'ākia* are said to be deadly to humans. One of the death penalties in old Hawai'i, which upheld a rigid set of laws known as the *kapu* system, was to force the guilty party to drink *'ākia* juice, causing seizure and death.

'AKOKO (ah-koh-koh)
Chamaesyce celastroides
Euphorbiaceae (Spurge Family)
Endemic

This shrubby member of the spurge family is highly variable, and can be found in a wide range of dry habitats, especially along the coast. This plant can grow as either a crawler or a small tree. The leaves, which appear in pairs on opposite sides of the stem, tend to fall off in the summer. The milky sap of the *'akoko* was given to lactating mothers as a supplement, and for weaning a child, the sap was mixed with *poi* or *kalo* leaves and given to the infant.

ʻĀKULIKULI (ah-coo-lee-coo-lee)
Sea Purslane
Sesuvium portulacastrum
Aizoaceae (Fig-Marigold Family)
Indigenous

This crawling succulent can be found in many coastal habitats on all the main islands. The stems are often tinged with red and the leaves are rounded and elongated. The herb has small flowers that are pale purple to whitish. Hawaiian lore mentions tiny islands with an inviting carpet of nothing but cool, soft *ʻākulikuli*.

'ALA'ALA WAI NUI WAHINE
(ah-lah-ah-lah vai-noo-ee wah-hee-neh)
Plectranthus parviflorus
Lamiaceae (Mint Family)
Indigenous

This is a common plant to find growing in the dry areas of each island, often hiding under trees or in shady cracks in the lava. The mint-looking leaves are somewhat fleshy, but not aromatic as one expects from a mint. This plant was used in a range of medicinal mixtures, usually utilizing the leaves in conjunction with other medicinal herbs such as *noni* and *'awa*.

'AUHUHU (ow-who-who)
Tephrosia purpurea
Fabaceae (Pea Family)
Polynesian-Introduced

This plant was introduced to Hawai'i by the Polynesians specifically for use as a fish poison, and was grown near the coast for easy gathering. The stem and seed pods of *'auhuhu* are crushed to form a poison that is effective only on fish, and causes no known reactions in mammals or birds. Hawaiians harvested their extensive saltwater and brackish water fishponds by chuming and then dumping the poisonous infusion into the water. The fish were temporarily paralyzed, making it easy to collect the largest ones while leaving the younger fish to fatten up.

HALA (hah-lah), Screwpine
Pandanus tectorius
Pandanaceae (Screwpine Family)
Polynesian-Introduced

The *hala* tree is easily recognized by its supporting base of aerial roots, leading to one of its nicknames: the walking tree. This mangrove-like growth pattern supports the top-heavy branches, and is an adaptation that allows *hala* to filter small amounts of salt water and survive at the coastline. The long green leaves are lined on both sides with sharp spines and grow like a fountain overflowing at the stem tips. At the center of this fountain is a large, circular fruit cluster that resembles a pineapple. The leaves, growth pattern, and fruits have tricked many visitors into thinking that this was indeed a pineapple, and lends to another nickname: the pineapple tree. These small fruits are edible, although very fibrous. When the meat rots away, the tuft of fibers that remains was used in ancient times as a natural paintbrush.

The most significant use of the *hala* tree is for weaving. The long, tough leaves are unparalleled among natural weaving materials for their combination of durability and pliability. Hats, mats, bags, and, most importantly, the sails of the Polynesian voyaging canoes are woven from *hala*. These leaves, when properly prepared, were strong enough to stand up to two-month, open-ocean voyages and land the Hawaiians here safely. Better yet, if the sails happened to rip, one simply needed to weave in a patch with some extra leaves!

HAU (how)
Hibiscus tiliaceus
Malvaceae (Mallow Family)
Indigenous(?)

This huge shrub can be found near waterways, springs, and other damp areas in lower elevations. The five-petaled flowers are red when young but turn bright yellow as they open, then fade to crimson before they fall from the plant. They seem to always land face up, and Hawaiian rivers near the ocean are often littered with hundreds of floating *hau* flowers, as if for some elaborate wedding. The trunk and branches of this tree twist and loop, forming impenetrable thickets that ancient peoples used around villages as a natural defensive barrier.

Hau was most appreciated, however, for its unparalleled ease in making natural cordage. The bark layer can be easily stripped from a *hau* branch and the inner bark fibers separate themselves into long, tough lengths of fiber, which were then twisted into cordage for a variety of uses. So prevalent was *hau* fiber use in old Hawai'i that the "grass skirts" many people associate with the islands are actually not grass at all, but the fibers of the *hau* bark! There is still some debate as to whether the Polynesians brought *hau* to Hawai'i or if it was already here. It is very probable that the answer is both!

HINAHINA (hee-nah-hee-nah)
Heliotropium anomalum
Boraginaceae (Borage Family)
Indigenous

These small shrubs can be found in the sand or rocks near the coast. The Hawaiian name means "silver," and refers to the silvery appearance of the *hinahina* caused by the soft hairs that coat the plant. The tiny white flowers grow in clusters and are quite fragrant. The leaves of the *hinahina* were used in a broad range of medicinal concoctions to treat illnesses such as thrush, asthma, and womb disorders.

'IHI (ee-hee)
Portulaca lutea, P. molokiniensis
Portulacaceae (Purslane Family)
Indigenous/Endemic & Rare

These two succulents are very closely related, and both grow in the low, dry coastal areas. *Portulaca molokiniensis* (pictured) developed in the isolation of Kaho'olawe island and the islets Molokini and Pu'uhoa'e. They can be differentiated by their growth habits, with *P. lutea* growing more prostrate and having more irregularly spaced and orientated leaves. The lemon yellow flowers of both species occur at the tips of the stems.

'ILIMA (ee-lee-mah)
Sida fallax
Malvaceae (Mallow Family)
Indigenous

These small shrubs can grow upright in the dry forest, but are often found as prostrate ground covers near the ocean. It is a highly variable species that can vary greatly in leaf size and shape, growth pattern, and color. The orange flowers are greatly prized in lei-making tradition, where they are known for attracting the mischievous spirits that lead to fun and adventure. *'Ilima* was so prevalent in the rocky and dry areas unsuitable to agriculture that those areas were referred to as *wao 'ilima*, or "realm of the *'ilima*."

KAMANI (kah-mah-nee)
Alexandrian Laurel
Calophyllum inophyllum
Clusiaceae (Mangosteen Family)
Polynesian-Introduced

This large, attractive tree is commonly found in landscapes around hotels and parks. *Kamani* has a sweet, maple-syrup-smelling sap that can be found in the deeply fissured bark. The bright yellow sap and dark purple bark make an attractive contrast. The large nuts were commonly hollowed out for whistles and other crafts. The hard wood of the *kamani* was popular for making Hawaiian calabashes due to its wide, squat trunk. It is a relatively dark wood, occasionally taking on a dark red tinge similar to that of the *kauila*. *Kamani* was traditionally considered the third most important wood source for carving, ranked behind *kou* and *milo*. The tree was also an important medicinal plant. The sap of the *kamani* tree was useful in treating ulcers and other stomach problems.

KAUNA'OA (cow-nah-oh-ah)
Native Dodder
Cuscuta sandwichiana
Cuscutaceae (Dodder Family)
Endemic

Kauna'oa is a parasitic plant that can be found on a variety of hosts in coastal and occasionally dry forest areas. The vine-like yellow or orange stems have no leaves, and support tiny whitish flowers. While juvenile plants may have roots, mature plants are not attached to the ground, rooting instead into its host plant. *Kauna'oa* was a favorite lei-making material in old Hawai'i and is the official flower of Lāna'i. *Kauna'oa* was used medicinally to treat such ailments as chest colds and post-childbirth symptoms for women.

KOU (koh)
Cordia subcordata
Boraginaceae (Borage Family)
Indigenous

These mid-sized trees often mark the sites of old Hawaiian coastal villages. The *kou* tree produces a profusion of small orange flowers, which were commonly strung for lei, and today can occasionally be found in coastal and dry forest habitats. Over 90% of the Hawaiian bowls in the Bishop Museum collection were crafted from the wood of the *kou* tree. It was a favorite for food platters and bowls because it is an attractive wood and, unlike the wood of *koa*, does not affect the flavor of the meal due to its low acid level. *Kou* wood is a beautiful dark brown, often with black streaks, and can be mistaken for *koa* to the untrained eye.

MAIAPILO (mai-ah-pee-loh)
Hawaiian Caper Bush
Capparis sandwichiana
Capparaceae (Caper Family)
Endemic, Vulnerable

This light green shrub can be found
growing on lava or soil along the dry
coasts. The plant supports gorgeous bright
white flowers, noted for their "waterfall"
of white stamens and heavily perfumed
fragrance. The flowers of the *maiapilo*
bloom shortly after sunset and tend to
wilt by the following mid-day. The intoxicating scent of the flower is almost overpowering
at night and fades rapidly after the morning. This shrub is a member of the caper family,
and is closely related to the species whose flower buds are turned into pickled capers.

MAKALOA (mah-kah-loh-ah)
Cyperus laevigatus
Cyperaceae (Sedge Family)
Indigenous

This sedge is found in most wetland habitats in Hawai'i. *Makaloa* was prized for producing the softest and finest woven mats in Hawai'i. The sedge is woven in elaborate patterns with as many as 30 strands in an inch. *Makaloa* can still commonly be found in the few Hawaiian wetlands that have not gone dry due to human activity, and is even used to treat wastewater in natural graywater treatment on Moloka'i.

MA'O (mah-oh)
Hawaiian Cotton
Gossypium
tomentosum
Malvaceae
(Mallow Family)
Endemic
Vulnerable

These silvery shrubs occur primarily on the rocky coastal plains of each island. *Ma'o* has a distinctive, three-lobed leaf and large, hibiscus-shaped flowers that are bright yellow. This cotton has relatively short fibers, and was therefore not useful for cordage or clothing. The hairs were gathered for tinder and used for medicinal application, much as cotton swabs or Q-tips are used today. The plant was commonly used to extract a green dye, a surprisingly difficult color to come by in natural dyes.

MILO (mee-loh)
Portia Tree
Thespesia populnea
Malvaceae (Mallow Family)
Indigenous(?)

In the same family as *Hibiscus*, these common coastal trees can be seen covered with beautiful yellow flowers. This is the most common native coastal tree, growing in sand or rock at shorelines around the islands. It has amazing growth rates in its younger years, producing a light, yellow wood. As *milo* reaches its maximum height of 25–30 feet, the wood becomes denser and darkens to a near-purple color. *Milo* wood was traditionally used for carving beautiful bowls and calabashes.

NANEA (nah-nay-ah)
Mohihihi, Beach Pea
Vigna marina
Fabaceae (Pea Family)
Indigenous

These vine-like plants usually form thick mats of ground cover in coastal areas. The large, three-parted leaves are supported by thick, almost succulent stems. The plant has a bright yellow flower. Its vigorous growth and nitrogen fixing properties make this a great mulching plant.

NAUPAKA KAHAKAI
(now-pah-kah kah-hah-kai)
Scaevola taccada
Goodeniaceae (Goodenia Family)
Indigenous

Naupaka kahakai is a common native coastal species, and this shrub can be found at virtually every beach habitat on the island. Its high tolerance for salt spray and ability to live directly on exposed lava rock has allowed it to remain in areas too harsh for competing species. The most striking feature of the *naupaka* bush is that the five flower petals are arranged in a semi-circular shape, giving the appearance of a "half-flower."

There are also several mountain species, known as *naupaka kuahiwi* (now-pah-kah coo-ah-hee-vee), that appear to have the matching half-flower to the ocean variety. This has led to many Hawaiian stories. One story tells about a princess and a commoner whose love could not be realized due to the strict laws regarding class. After the two of them committed suicide together (or were killed when their love was discovered, depending on which version of the story you hear), the gods continued the banishment of their love by placing one by the ocean and the other in the mountains, keeping apart forever their two half-flowers.

NEHE (neh-heh)
Melanthera integrifolia
Asteraceae (Sunflower Family)
Endemic

This light green to silvery ground cover can be
found in the sun-baked strands and coastal areas
as small herbs, but grows into dense mats when
found near water sources. A sunflower relative,
the *nehe* boasts beautiful bright yellow, daisy-like
flowers. Like many coastal plants, the hairs of the
nehe reflect the sunlight to produce a silvery color.

NIU (nee-oo), Coconut
Cocos nucifera
Arecaceae (Palm Family)
Polynesian-Introduced

Coconuts are perhaps the most useful plant on earth. Starting from the bottom, the roots can be divided and braided into strong cordage. The tree trunk was used primarily for large *pahu*, or drums, but could be hollowed out to make a small canoe or used as a house post. The crown can also be harvested for the edible heart of palm. The leaves are woven to make hats, mats, baskets, fans, and hundreds of other items by many cultures around the world. The leaf midribs were used as needles in old Hawai'i.

The large nuts are sheathed in a thick husk, which can be braided into durable twine. Within the husk is the rock-hard shell of the nut, which was commonly used to make beautiful bowls and spoons. The hard nut is relatively hollow, containing mostly "coconut milk" and a small layer of fruit, both of which are edible. Copra, or palm oil, can be extracted from the oily fruits. Without these natural watertight containers of food and drinkable liquid, it is doubtful that the ancient Polynesians would have been able to make the long ocean voyages that they did.

The Polynesians considered *niu* a resource of the highest level and made the plant sacred to Kū, the primary masculine symbol in ancient Hawai'i. To destroy an enemy's coconut trees was considered the gravest of insults and a declaration of war. In fact, the conquests of Kamehameha the Great, the first monarch to unify the Hawaiian archipelago under one ruler, were started in just this manner. This battle was waged directly below the Garden, in the coastal plains of Moku'ōhai, just south of Kealakekua Bay. The battle started nearly 20 years of war before Kamehameha fulfilled the prophecies foretold at his birth and unified the Kingdom of Hawai'i.

NONI (noh-nee)
Indian Mulberry
Morinda citrifolia
Rubiaceae (Coffee Family)
Polynesian-Introduced

These small trees are easily recognized by their large, oval, glossy leaves, which were very important in Polynesian medicine. The leaves were used to cure most types of muscle and joint pain by placing the leaves directly on the infected area and using heated stones and massage to release the wax into the skin. Today, *noni* fruit has become very popular in holistic and herbal medicines across the world. The fruit has been praised as a "cure all" medicine, used for everything from asthma to cancer. Contrary to popular belief, there is no evidence to support Hawaiians ingesting *noni* fruit except in times of extreme famine. Take a sniff of a clear, overripe fruit and find out why!

All five stages of the *noni* fruit can typically be found on one branch. The fruit starts as a green bud, which supports several small white flowers in the second stage. Where the flowers fall off they leave scars, or eyes, on what is now a strange, knobby green mass. In the fourth stage this mass smooths out to form a green fruit, and the final stage is a ripe, yellow/white fruit with apparent eyes. When the fruit is overripe, it turns into a clear, jelly-like substance and acquires its musty scent.

ʻOHAI (oh-hai)
Sesbania tomentosa
Fabaceae (Pea Family)
Endemic
Endangered

ʻOhai is a short shrub with leaves that vary between light green and completely silver. The leaves are sweet smelling, and the more silvery the leaves the more fragrant they tend to be. The softness of the leaves comes from thousands of tiny hairs that cover the plant, which acts as a natural sunscreen to shade the ʻohai from the sun as well as help gather small moisture particles out of the air. ʻOhai was highly prized for its flowers, which were used extensively in lei-making.

'OHE MAKAI (oh-hay mah-kai)
Reynoldsia sandwicensis
Araliaceae (Ginseng Family)
Endemic, Rare

These tall and usually straight trees can be found in the dry forests of each island. The spreading crown of the tree can be as much as 90 feet high, and the light green leaflets often stand out against the drier background. The wood was traditionally used to make stilts. Wood from a particular grove on Moloka'i was reputed to be poisonous, and was used for making poison images—images used by Hawaiian sorcerers to bring ill will upon others. This is the tree form of the goddess, Kapo.

PĀʻŪ O HIʻIAKA
(pah-oo oh hee-ee-ah-kah)
Jacquemontia ovalifolia ssp. *sandwicensis*
Convolvulaceae (Morning Glory Family)
Endemic

This lovely crawling vine has silvery leaves and small, pale blue flowers. The plant was given its name, *pāʻū o Hiʻiaka*, from Hawaiian mythology. The volcano goddess, Pele, sent her younger sister, Hiʻiaka, on an epic journey across the islands to bring back a lover whom Pele had met. To aid in the difficult quest, Hiʻiaka was granted a skirt that possessed magical powers, and it was tied with the beautiful lavender-flowered vine that has since been known as "the skirt (*pāʻū*) of Hiʻiaka."

PILI (pee-lee)
Heteropogon contortus
Poaceae (Grass Family)
Indigenous(?)

This tall clump grass, up to three feet high, can occasionally be found in dry areas where invasive grasses, such as fountain grass, now typically dominate the landscape. This grass may have been introduced by the Polynesians, who favored the *pili* due to its pleasant odor. The grass was used extensively for thatching houses, but also for stuffing mattresses, padding floors, and as a tinder. *Pili* can be recognized by its kinked, or contorted, seed spikelets, which angle downward instead of being erect like the spikelets of most grasses.

PŌHINAHINA (poh-hee-nah-hee-nah)
Beach Vitex
Vitex rotundifolia
Verbenaceae (Verbena Family)
Indigenous

These crawling shrubs grow in primarily strand areas, with short erect branches that rarely extend more than a foot. The leaves are light green to grayish in color and have a faint sage-like smell when crushed. The small purple flowers occur in clusters on small stalks. This species was used medicinally, primarily in concoctions with other plant species. It is now a popular ground cover for landscapes.

PŌHUEHUE (poh-who-eh-who-eh)
Beach Morning Glory
Ipomoea pes-caprae ssp. *brasiliensis*
Convolvulaceae (Morning Glory Family)
Indigenous

These vines can be found growing over rocks or sand in the strand habitat. Its ovate, light green leaves often begin to fold themselves in half, probably to protect themselves against the intense sunlight and heat. The beautiful flowers vary from light purple to pink, and typically open in the morning. The long vines often made a handy rope when at the ocean. *Pōhuehue*, as many other morning glories, was used for a variety of medicines, particularly intestinal and colonic cleanses.

PUA KALA (poo-ah kah-lah)
Hawaiian Poppy
Argemone glauca
Papaveraceae (Poppy Family)
Endemic

This silvery plant is an extremely hardy poppy that grows throughout the coastal and dry forest areas, on the leeward sides of all the major islands from sea level to 6,000 feet. The plants are often blooming with their delicate white flowers. *Pua kala*, which literally means "thorny flower," is covered in sharp, stiff pickles on the stalk, leaves, and seed pods. *Pua kala* was valued highly for its medicinal properties, as are many poppy species around the world. The bright yellow sap of the plant was applied to warts and other skin disorders, while brews made from the seeds and sap acted as a narcotic and analgesic, particularly for oral surgeries.

'ŪLEI (oo-lay)
Osteomeles anthyllidifolia
Rosaceae (Rose Family)
Indigenous

This plant is usually a crawling shrub, but occasionally will grow upright and produce a small tree. The wood is extremely flexible and springy, and was utilized by the Hawaiians for making bows and arrows. Although it is not well known, the Hawaiians commonly used bow and arrows to hunt birds and rats. The springy wood of the *'ūlei* could be bent into circles for fish net hoops and was also used to make a variety of children's toys. The small berries turn white with purple dots when ripe and are edible and occasionally tasty, although usually grainy.

DRY FOREST ZONE

DRY FOREST HABITATS are usually found on the leeward side of each island, and extend from the coastal zone (approximately 500 feet elevation) until the rainfall boundary of 150 centimeters (59 inches) is reached, which varies in elevation from place to place and can reach elevations of 6,000 to 7,000 feet.

The dry forest habitat is the most threatened of all Hawaiian habitats, and is home to many rare and endangered species. The destruction of the dry forest started with the Hawaiians, who cleared vast amounts of this landscape for cultivation. These changes were amplified with the arrival of Westerners, who clear-cut even more of the dry forest for cattle and sugarcane. Today the dry forest faces a plethora of threats. Introduced feral animals such as goats, cattle, donkeys, deer, and sheep eat the young trees and prevent new generations from being established in the forest. Invasive plants also present a major threat, such as the fire-adapted fountain grass that annually spreads wildfires in and around the dry forests. Perhaps the most complete threat the dry forest faces is urbanization, as people prefer to live in the warm and dry areas that provide the dry forest habitat.

A continuous canopied forest once extended from North Kohala to South Kona before humans began to alter the landscape. Within this forest lived hundreds of now-extinct species. An important habitat on the Big Island is the forest margin, where lava flows have created what are known as fringe habitats. The new lava flows allow shrubs and other plants to thrive before being shaded out by the larger trees.

The plants of the dry forest are often deciduous, and grow slowly. *Lama*, one of the primary dry forest trees may add only one millimeter ($\frac{1}{16}''$) of girth per year. In monarchal Hawai'i, the taking of these trees was regulated by the chiefs and involved a considerable amount of ceremony, which greatly helped to slow the harvest rate of these precious trees. Today ancient dry forest trees are often pushed over by bulldozers without a second thought.

ʻAʻALIʻI (ah-ah-lee-ee)
Dodonaea viscosa
Sapindaceae (Soapberry Family)
Indigenous

ʻAʻaliʻi is one of the first plants to re-inhabit a new lava flow, and can be found in a variety of drier habitats from the coast to the mountains. Its bright red seed capsules are primarily filled with air and have little wings that allow them to be caught up in the breeze and carried onto uninhabited areas. The red seed capsules are a favorite in woven lei, known as *haku*, which commonly adorned hula dancers. Unlike most trees in Hawaiʻi, ʻaʻaliʻi has a deep tap-root, which allows it to tap into deeper water supplies and stand up to high wind levels.

'AIEA (eye-eh-ah)
Nothocestrum breviflorum
Solanaceae (Nightshade Family)
Endangered

Today these stout trees can only be found in the dry forests of Hawai'i Island. The tree flowers profusely with many yellow-green flowers that are known for their sweet, strong scent. This plant received additional conservation attention when the Blackburn Sphinx Moth (*Manduca blackburni*), the first Hawaiian insect to be put on the endangered species list, was found to host on the 'aiea. 'Aiea is the only known native plant that the large, horned caterpillars, which can be over 2 inches long, are known to eat. The town of 'Aiea on O'ahu is named for a closely-related species.

'AKOKO (ah-koh-koh)
Chamaesyce olowaluana
Euphorbiaceae (Spurge Family)
Endemic
Vulnerable

These sparse trees can only rarely be seen in dry forests on Hawai'i and Maui. The small elongated leaves are usually clustered at the canopy, and the pinkish flowers are inconspicuous, but often plentiful. The sap from many of the *akoko* species was used in a number of medicinal concoctions, and is said to increase the amount of milk produced by a nursing mother.

ALAHE'E (ah-lah-heh-eh)
Psydrax odorata
Rubiaceae (Coffee Family)
Indigenous

Alahe'e was very common in dry forests of old, comprising a majority of the sub-canopy, and is still relatively common in Hawaiian dry forests. It can be recognized by its shiny, dark green leaves, which do not lose their luster even in the harshest of droughts. The dense wood of the *alahe'e*'s straight branches was used extensively to make a traditional Hawaiian farming tool, the *'ō'ō*. This was the Hawaiian hoe, shovel, and crowbar all in one, and was the primary farming tool used. The *alahe'e* was the proper mixture between density and weight, which was necessary so the *'ō'ō* did not break or tire its user quickly.

ʻĀWEOWEO (ah-veh-oh-veh-oh)
Chenopodium oahuense
Chenopodiaceae (Goosefoot Family)
Endemic

This plant can be found in a variety of dry forest habitats, and is somewhat variable among them. Leaves can be light green or bright silver, and the plant itself can grow as either an erect or spreading shrub. ʻĀweoweo shares its Hawaiian name with a good-eating fish in Hawaiʻi. The leaves of the plant, when crushed, have an undeniably fishy smell. The fresh leaves of the plant were cooked and added to other vegetables and starches for their fishy taste.

HALA PEPE (hah-lah peh-peh)
Pleomele hawaiiensis
Agavaceae (Agave Family)
Endemic
Endangered

Hala pepe is a slow-growing
member of the Agave family that prefers hot and dry, almost
desert-like climates. The leaves cluster at the top of the erect
branches, and the trunk and branches are ringed with scars
where old leaves have fallen off. The plant goes through one
flowering display each year, when hundreds of cream yellow
flowers appear in long, drooping clusters. After a time the
flowers are replaced by red berries, which can be seen on
the plant for most of the year. The long, skinny leaves were
occasionally used for hula skirts and adornment, and this
tree is one of the five plants that were sacred to the hula altar.

HAU KUAHIWI (how coo-ah-hee-vee)
Hibiscadelphus hualalaiensis
Malvaceae (Mallow Family)
Endemic
Endangered

This is an extremely rare species that grows only on the dry slopes of Hualālai Volcano. *Hau kuahiwi*, which actually refers to all seven endemic species of the *Hibiscadelphus* genus, is believed to have co-evolved with native species of finches, known as honeycreepers. The curved, tubular flowers of the genus are presumably adapted to fit the beak of the native honeycreepers for pollination. Because of the difference in habitats that each species occurs in, it is very possible that each of the seven species of *Hibiscadelphus* had its own species of honeycreeper that would pollinate its flowers.

'ILIE'E (ee-lee-eh-eh)
Leadwort
Plumbago zeylanica
Plumbaginaceae (Plumbago Family)
Indigenous

This thick ground cover is the only Plumbago family member native to Hawai'i, easily recognized by the sticky green mass at the base of each flower. *'Ilie'e* is occasionally referred to in Hawai'i as the "poor girl's earrings," since the flowers can be easily adhered to one's ears and worn around as decoration. The dark, sticky sap of the *'ilie'e* was used to darken tattoos in old Hawai'i.

KAUILA (cow-ee-lah)
Alphitonia ponderosa
Rhamnaceae (Buckthorn Family)
Endemic

Two related rare trees, both called *kauila,* share botanical and ethnobotanical qualities. The dull leaves and small flowers make them difficult plants to spot in the wild. They have a reputation as being the hardest of all Hawai'i woods, and will sink in water due to their density. The wood was consequently used by Hawaiians to make their most durable items, such as working implements and weapons.

KOAIʻA (koh-eye-ah)
Acacia koaia
Fabaceae (Pea Family)
Endemic
Rare

This small, gnarled tree can be found growing in upper elevation dry forests of Hawaiʻi and Maui. It is easily recognized by its sickle-shaped leaves, and is the dryland cousin of *koa*, the dominant wet forest species in Hawaiʻi. The wood of the *koaiʻa* is much harder and denser than *koa*, and was a favorite for making spears, canoe paddles, and *kapa* beaters.

KOALI (koh-ah-lee)
Morning Glory
Ipomoea spp.
Convolvulaceae
(Morning Glory Family)
Indigenous

There are several morning glory species in Hawai'i that can be found in a variety of dry to mesic habitats. The heart-shaped leaves vary in size, and the flowers can be an array of white, pinks, and purples. There is still some debate regarding whether some of the species are truly indigenous or whether they were introduced by the early Hawaiians, who used the plants extensively in a variety of medicines.

KOKI'O (koh-kee-oh)
Kokia drynarioides
Malvaceae (Mallow Family)
Endemic
Endangered

This extremely endangered hibiscus species has an amazing survival story. Pressures on its habitat from human development, wildfires, and feral goats reduced this plant to only four remaining individuals in the wild. After receiving federal permission, botanists supplied the Amy Greenwell Garden with seeds nearly 15 years ago. Since then the Garden has grown thousands of seedlings, primarily supplied to managed conservation efforts on the island, such as Ka'ūpūlehu Dry

Forest and Pu'uwa'awa'a Natural Area Reserve, both located north of Kona. We have grown so many *koki'o* that we now even sell them to home consumers for private landscaping.

The red flowers show unique curvatures and are a popular subject for local artists. The bright red, viscous sap of the plant dries into a nearly waterproof resin that was used by Hawaiians to dye their fishing nets. The resin increased the lifespan of the nets, and the red coloring allowed them to catch more fish. This is because red light is filtered underwater, making red objects more difficult for fish to see. *Kokia cookei*, an even more endangered species, was also found in the dry forest and can be distinguished by its relatively bare branches with "pom-pom" leaf clusters at the tips.

KOLOMONA (koh-loh-moh-nah)
Senna gaudichaudii
Fabaceae (Pea Family)
Indigenous

This member of the pea family is found on fringe habitats and young lava flows in sunny areas. This rather common plant played an important role in the dry forests. As a member of the pea family, *kolomona* is able to pull nitrogen out of the air, which is a rare ability, and is an important source of that critical nutrient for plants.

KO'OKO'OLAU (koh-oh-koh-oh-lau)
Bidens spp.
Asteraceae (Sunflower Family)
Endemic

There are 23 species of *Bidens* in Hawai'i,
19 of which are endemic to the islands. There
are several species represented at the Garden,
including *B. micrantha*, *B. hawaiensis*, and *B.
menziesii*. These plants are close relatives to a
common weed called the Spanish needle, noted
for its long black seeds which stick to clothing.
The leaves, buds, and flowers of the *ko'oko'olau*
were brewed into teas that were used to treat
weight loss, thrush, especially in children, and
other ailments.

KO'OLOA 'ULA (koh-oh-loh-ah oo-lah)
Abutilon menziesii
Malvaceae (Mallow Family)
Endemic
Endangered

These attractive silvery shrubs are in the Hibiscus family and boast beautiful, although small, red hibiscus-type flowers that hang facing the ground. On the Big Island, the last wild population of this very rare plant, located in the Puakō area, disappeared several years ago. Now on Hawai'i Island, *ko'oloa 'ula* exists only in managed wildlife preserves where it has been outplanted. In general the Mallow family is very susceptible to insects, drought, and disease, and consequently over half of the 25 or so native species are on the endangered species list. The Garden houses two other Abutilons, both of which are extremely endangered: *Abutilon eremitopetalum*, which means "hidden-petaled Abutilon" in Latin, and *Abutilon sandwicense*, whose large green and red flowers are well worth seeing. Both species have the characteristic of downward-facing flowers.

KULU'Ī (coo-loo-ee)
Nototrichium sandwicense
Amaranthaceae (Amaranth Family)
Endemic

This shrub can weather the worst of droughts, and is found throughout dry habitats. The silvery color comes from thousands of tiny hairs that coat the plant, and the variable leaves are smaller in drier climates. The downy flower spikes and wood of *kulu'i* were used as a form of ancient fireworks. The material was packed into a hollow stem, such as that of bamboo. Then a fireworks specialist would climb to the top of particular cliffs where the winds were just right. The firework was lit and thrown, and the wind would ignite the flower spikes, which were ejected from the bamboo in a spectacular display of aerial embers.

LAMA (lah-mah)
Diospyros sandwicensis
Ebenaceae (Ebony Family)
Endemic

These trees were the dominant
dry forest canopy along with the
'ōhi'a, and can still be found today
in most dry forest habitats. The
new growth of the plant is reddish,
and fades to a lighter, then darker,
green. The flowers are extremely
inconspicuous, but lead to the
development of bright red fruits up
to an inch long. In the same family
as persimmon, the fruits are sweet
and delicious. *Lama*, in the ebony
family, has a beautiful wood that
was held sacred to the goddess of
hula, Laka.

LOULU (loh-oo-loo)
Pritchardia spp.
Arecaceae (Palm Family)
Endemic

Pritchardia is the only genus of palms native to Hawai'i. Referred to collectively by the Hawaiians as *loulu*, there are 19 different Hawaiian species of *Pritchardia,* nine of which are endangered or threatened. These fan palms could traditionally be found in many of the Hawaiian habitats, but today are located most often in dry forest regions. The Garden

houses several different species of *loulu,* which can be distinguished by differences in plant height, trunk width, and leaf shape. The wood was used for house posts, altars, drums, and canoes, and the leaves were woven into various hats and baskets. The tiny palm fruits are not very edible nor do they have the oil content of their cousins, the coconut.

The Garden is a participating institution in the Center for Plant Conservation, and houses nine accessions to the National Collection, including *P. affinis.* This means that we are primarily responsible for assuring the survival of live plants and propagative material until each plant has been removed from the endangered species list. Since our appointment, we have grown thousands of *P. affinis* that have been planted, sold, or given away to aid in its recovery.

Other species of *loulu* found on the property include *P. beccariana,* an erect palm that can reach over 60 feet in height; *P. glabrata,* a miniature palm up to 4 feet tall; *P. napaliensis,* a 12-foot palm with a very slender trunk; and *P. schattaueri,* a large palm that can reach heights of over 100 feet.

MA'O HAU HELE (mah-oh how heh-leh)
Hibiscus brackenridgei
Malvaceae (Mallow Family)
Endemic
Endangered

Ma'o hau hele is a large dry forest shrub that can form dense, almost impenetrable, thickets much like the *hau* tree. The Hawaiians must have seen the relationship between these two very different hibiscus relatives, for the name *ma'o hau hele* literally means "green traveling *hau*." This plant, which once covered huge areas of dryland habitat on all the major islands, is now highly endangered, found more often in backyards than in the few remaining Hawaiian dry forests. This is the official state flower, the yellow hibiscus. Unfortunately, due to the rarity of this beautiful plant, many people believe that the red Chinese hibiscus is the state flower. An attractive yellow-green dye is made from the leaves and flowers.

'ŌHI'A (oh-hee-ah)
Metrosideros polymorpha
Myrtaceae (Myrtle Family)
Endemic

The species name (polymorpha) of the *'ōhi'a* means "many formed," in reference to the widely varying characteristics of the tree as it grows in nearly all of the diverse habitats that Hawai'i has to offer. *'Ōhi'a* can be found as 100-foot tall trees in the rainforest or as 1,000-year-old bonsai trees in the mountain bogs. The light green, ovate leaves found on the very summit of Hualālai look nothing like the silvery, elongated ones found on the newest of lava flows. Almost everywhere you go in Hawai'i there will be *'ōhi'a*...if you can recognize it. The only stable characteristic of the plant is its pincushion-type flower, which varies from red, yellow, orange, to pink.

'Ōhi'a is known as a pioneer species. Its tiny, spore-like seeds are carried by the wind to settle everywhere on the island. It is the first tree to appear on new lava flows, helping to create livable habitats for other species. It also forms or contributes to the canopy of virtually all Hawaiian forests—wet or dry, high or low—and is home to and food for many forest birds and insects. The 'ōhi'a tree is sacred to the volcano goddess Pele, and the bright red blossoms are reminiscent of a volcanic eruption. Indeed the 'ōhi'a not only survives, but seems to thrive in the sulfur-filled air near the volcanic vents of Pele.

PŌPOLO (poh-poh-loh)
Solanum americanum
Solanaceae (Nightshade Family)
Indigenous(?)

This is a common plant in a variety of wet and dry habitats. Known in Hawai'i as *pōpolo* berry, it is not known whether this plant was indigenous to Hawai'i, or if it was a post-contact introduction. There are also several endemic species of *Solanum*, most of which are very rare. Although considered a weed by today's standards, the different *pōpolo* species are valuable medicinal herbs in traditional Hawaiian culture, and *pōpolo* is still used extensively by practitioners today for a wide variety of ailments and also for external injuries.

'UHALOA (oo-hah-loh-ah)
Waltheria indica
Sterculiaceae (Cacao Family)
Indigenous(?)

'Uhaloa is still a common dryland herb and can be found in dry, open habitats.
The Hawaiians, like all indigenous cultures, relied on plants for their medicines.
A popular method of traditional medicine preparation was the brewing of medicinal
teas. Some teas were taken regularly to soothe weary muscles or aid in general health,
while other teas were brewed for specific remedies. The dense flower buds of the
'uhaloa were brewed and the tea used primarily to treat throat ailments such as
asthma and tonsillitis. It was also used as a remedy for chest pains, getting rid of
that run-down feeling, and for weight loss.

UHIUHI (oo-hee-oo-hee)
Caesalpinia kavaiensis
Fabaceae (Pea Family)
Endemic
Endangered

There are less than 100 remaining *uhiuhi* trees in the wild, found in the dry forest of the larger islands. The light green canopy can grow up to 30 feet tall and the rough bark is dark and often fractured. The bright pink and purple flowers are about an inch long, and grow on spikes containing several dozen flowers. The flat, rounded seed pods are visible on the plant for most of the year. *Uhiuhi* wood, black in color, is extremely dense and will sink in water. It was used by Hawaiians for spears, farming and fishing tools, and occasionally, building.

POLYNESIAN-INTRODUCED CROPS

WHEN THE POLYNESIANS first arrived on the various islands of the Pacific they found them to be virtually devoid of edible plants, and Hawaiʻi was no exception. There were only a few trees and shrubs that produced small edible berries. The Polynesians, with over a millennium of seafaring experience, knew that they would have to take along their own crops in order to survive. They discovered the best ways to transport each plant and developed techniques for protecting them from the deadly salt water on their long ocean voyages. They brought along plants by stem cuttings, root stalks, and tubers, but very rarely by seeds.

The most prominent rock features found in the Garden are the archaeological remains of the intensive Hawaiian agricultural system that covered 50 square miles of the uplands above Kona. Hawaiian farmers removed rocks from their planting areas and arranged them into low, mounded walls called *kuaiwi*, which ran down-slope. Between the *kuaiwi* were planted primary crops such as *kalo* or *ʻuala*, and along the edges of the *kuaiwi* were planted secondary crops, such as banana and sugarcane. It is thought that these taller border crops created shade, captured and preserved moisture, served as windbreaks, and provided mulch for the primary crops. The walls themselves were used as pathways to navigate through the elaborate field system.

Ahupuaʻa (ah-who-poo-ah-ah) literally means "pig altar," and refers to the rock markers that lined district boundaries. These rock altars were adorned with some representation of a pig, most commonly in the form of a carved block of *kukui* wood. In Hawaiian mythology, there existed a large, six-eyed, six-tusked pig god named Kamapuaʻa, whose constant rooting carved the deep valleys of the islands. When the need arose Kamapuaʻa could take the form of a pig, a triggerfish, or the *kukui* tree. If you look at the five-pointed leaf of the *kukui* you can still see the two ears and snout of Kamapuaʻa. While the altar within the Garden is a replication constructed by a volunteer, the majority of the rock work you will notice throughout the Garden are archaeological remains of the traditional Hawaiian farming system that existed here 400 years ago.

The term *ahupua'a* applies to the districts themselves as well. A typical *ahupua'a* extended from high in the mountain to the ocean, encompassing the broad range of climates and resources that the land had to offer. The ocean offered fishing, gathering, and recreation, and the coast was the primary living zone because of the low rainfall and pleasant conditions. Above the living zone the dry *kula* plains allowed for thatch and other materials. Moving up the mountain, rainfall increases and above the *kula* plains began the agricultural fields of the Hawaiians, which covered virtually all of the arable land. Where the agriculture stopped, the forest began, offering a range of resources such as timber, birds, and medicine. Above the forest, the summit of the mountain was a place for spirituality and learning such skills as navigation. These strips of land, therefore, offered everything to the people that they needed to survive.

'AWA (ah-vah)
Kavakava
Piper methysticum
Piperaceae (Pepper Family)
Polynesian-Introduced

'Awa was an important medicine and was also used extensively in religious and spiritual ceremonies. The plant takes about three years to mature, at which point the root is harvested, pounded with water, and strained into a strong drink. This drink exhibits narcotic properties, including heightened senses and dulled pain. In fact, the scientific name literally means "intoxicating pepper." *'Awa* was frequently drunk by the commoners, primarily to ease weary muscles and aid in sleep. The *ali'i*, or chiefs, also drank *'awa* for recreational purposes and would consume large quantities as a show of power.

The plant was also used extensively by a third class of Hawaiians, the *kāhuna*. *Kāhuna* are vaguely defined as "specialists" in various fields and with various rankings. Their fields ranged from war to healing, from divination to the natural sciences. Some *kāhuna* ranked higher than lesser *ali'i*, and some were *ali'i* themselves. Depending on their specialty, *kāhuna* would use *'awa* in varying quantities. For example, one might consume large amounts of strong *'awa* to evoke spiritual power for divination, or take small amounts of *'awa* to hone the senses and prepare for battle.

'AWAPUHI (ah-vah-poo-hee)
Shampoo Ginger
Zingiber zerumbet
Zingiberaceae (Ginger Family)
Polynesian-Introduced

'Awapuhi can today be found growing wild in many wet habitats. This ginger gets its nickname, the shampoo ginger, because of the thick sap contained in the ripe flower buds that is popular as a leave-in conditioner, and extracted for use in many hair care products. *'Awapuhi* can also be used as a flavoring in cooking and, like all the gingers, in a range of medicines.

KALO (kah-loh)
Taro
Colocasia esculenta
Araceae (Philodendron Family)
Polynesian-Introduced

Taro is indisputably the most important plant in Hawaiian culture. Besides being the staple food source, it was also the center of spiritualism, mythology, and social structure. In Hawaiian lore, the *kalo* plant is literally the elder brother of mankind. As the elder brother, *kalo* cares for and nourishes mankind, and we, in return, must respect and care for the *kalo*.

The *kalo* plant is high in carbohydrates, vitamin A, B, and C, and a variety of other elements and enzymes that are important to healthy living. *Kalo* contains calcium oxalate crystals in the leaves and tubers, and traditionally had to be steamed for over an hour before the crystals broke down and the plant became edible. Steamed *kalo* was then mashed into *poi*, which was the staple food of old Hawai'i. Breakfast, lunch, and dinner usually centered on *poi*, and the signal to begin eating was the uncovering of the *poi* bowl.

Kalo is typically cloned, with each progressive crop being derived from the last harvest. This method allows for farmers to keep the best varieties of *kalo*, while allowing the less productive ones to die off. The skilled Hawaiian farmers recognized over 300 varieties of *kalo*, which arose from mutations of the growing buds. Today, Amy Greenwell Garden has one of the oldest and most complete collections of Hawaiian *kalo* varieties, with our total count at over 70 different identifiable varieties.

KĪ (key)
Ti Leaf
Cordyline fruticosa
Agavaceae (Agave Family)
Polynesian-Introduced

This Agave family member can be seen throughout the islands, and in the wild is often a sign of traditional Hawaiian settlements. The Polynesians brought the stalk cuttings of this plant in their canoes, as any section of a branch will root and grow into a new plant very easily. The Hawaiians would steam and eat the large, tuberous root and derived many medicines from the sap and root. The most useful aspect of the *kī* was its long, broad, waxy leaves. These leaves were utilized in a number of ways. They could be twisted into the most basic of Hawaiian cordage or strung into a hula skirt or a raincoat. They were creatively bound to the feet to make disposable shoes, handy for walking across the barren lava fields of Hawai'i. Bundles of leaves could be used to thatch a house or make a broom. *Kī* leaves padded houses and trails, were laid over soil as a mulch to block out sun and suppress weeds, and used as plates during meals.

KŌ (koh)
Sugarcane
Saccharum officinarum
Poaceae (Grass Family)
Polynesian-Introduced

Sugarcane was a popular item in old Hawaiʻi, and is still enjoyed today. Hawaiians extracted the juice to sweeten dishes and medicines, or simply chewed on the raw stalk as a quick snack. *Kō* is very fibrous internally, and as it was chewed, it scrubbed the teeth and gums clean. The Garden displays many different varieties in our "Hall of Sugar," 41 in all. They vary widely in stalk color and pattern as well as in sugar content and taste. For nearly 100 years, sugar was Hawaiʻi's primary industry and export, and is responsible for the ethnic diversity of the islands today.

KUKUI (coo-coo-ee)
Candlenut Tree
Aleurites moluccana
Euphorbiaceae (Spurge Family)
Polynesian-Introduced

This is the state tree of Hawai'i, named so for its cultural importance as a resource. *Kukui* is a quick-growing, handsome tree that maintains a nice, round canopy.

It can be easily recognized by its light green, often silvery leaves that at maturity have five distinct points like the leaf of a maple tree. The tree fruits profusely year-round, producing up to 100 pounds of *kukui* nuts each year. The small green fruit contains hard-shelled seeds that were polished into a wide variety of ornaments and are still a popular lei item today. The nuts of the *kukui* are 80% oil and were an important source of lighting fuel in traditional times, leading to the tree's common name, "candlenut tree."

Kukui oil is still used for wood polishing and preservation, for lubrication, and for skin care and tanning lotions. The oily nature of the nut makes it an effective, although tasty, laxative. Finally, various dyes were extracted from the leaves, fruits, roots, and bark of the *kukui* tree, all of which contain tannin, a substance that makes dyes bind better to a variety of materials. It was the dark black ink made from the soot of burnt *kukui* nuts that the Hawaiians and other Polynesians used for their original art form: human tattooing!

MAI'A (mai-ah)
Banana
Musa spp.
Musaceae (Banana Family)
Polynesian-Introduced

Bananas have been cultivated since ancient times and have been an important food source for many cultures. Bananas have been grown for so long that human-selected traits have become the standard; for example, we are used to bananas not having seeds. Hawaiians cultivated many varieties of bananas. The most interesting of all might be what the Hawaiians called *mai'a hāpai*, which means "the pregnant banana." The fruit of the banana actually grows inside the stalk! The stalk gets fatter and fatter until someone comes along at the right time and cuts open the stalk to remove the bananas.

'ŌLENA (oh-leh-nah)
Turmeric
Curcuma longa
Zingiberaceae (Ginger Family)
Polynesian-Introduced

Turmeric is an annual herb in the ginger family, known best as a flavoring in Indian curries. However, the medicinal properties of the plant are truly precious. The rhizomes (underground stems) are bright orange and are the primary ingredient in many remedies. The rhizomes exhibit anti-inflammatory, anti-fungal, and anti-bacterial properties and were used to treat infections and swelling. It was also considered particularly effective against inner ear, throat, and nasal congestions.

A tea brewed from the rhizome was used as an immunity booster, stomach settler, and blood purifier. The orange color of *'ōlena* rhizomes also makes a yellow dye, which the Hawaiians used to dye their *kapa* for "work clothes." The multi-colored flowers are attractive for the short time they appear, resembling other coned flowers in the ginger family.

PIA (pee-ah)
Polynesian Arrowroot
Tacca leontopetaloides
Taccaceae (Tacca Family)
Polynesian-Introduced

The Hawaiians grew many root crops for food and medicine. This high percentage of crops grown from tuberous stems and roots may be in response to the long ocean voyages of the Polynesian people. Tubers store their own food and water and could survive long ocean journeys that would kill most stem cuttings and seeds. Root crops are also generally easy to grow and require minimal maintenance, making them ideal in an agricultural society.

The *pia* plant is an attractive annual with long stamens extending from its white flowers. *Pia* is a form of starch, and its use shows some of the culinary diversity of the Hawaiians. *Pia* tubers were grated, soaked, dried, and ground to make a starchy flour. This powdered

pia was mixed with coconut milk and sugarcane juice, and then baked in the *imu*, or underground oven. The result was a firm, sweet pudding called *haupia*, which is still popularly served today. Alternatively, the starch could be mixed with water to form a glue that was used to repair or layer *kapa* cloth.

'UALA (oo-ah-lah)
Sweet Potato
Ipomoea batatas
Convolvulaceae
(Morning Glory Family)
Polynesian-Introduced

The sweet potato was an embodiment of Lono, the god of rain and agriculture (among other things). The sweet potato poses a great question because it is the only Polynesian-introduced plant that does not originate in Southeast Asia. In fact, it hails from the opposite side of the Pacific Ocean in South America. There are several competing theories about the sweet potato's arrival in Polynesia: that the *'uala* floated here on a log, that the Polynesians sailed to South America and came back with the *'uala*, or that the South American Indians sailed into the Pacific, taking the *'uala* with them. However it arrived in Hawai'i, the sweet potato gained a strong importance here. The Hawaiian Islands are the tallest of the Polynesian islands. The semi-desert and desert habitats of the leeward mountain slopes were too dry to support large populations on taro, which required additional water. The sweet potato, however, thrived in these arid conditions and allowed the Hawaiians to increase the size of their dry-land settlements. There were over 50 varieties of Hawaiian sweet potato, 14 of which are represented in the Garden.

'ULU (oo-loo)
Breadfruit
Artocarpus altilis
Moraceae (Mulberry Family)
Polynesian-Introduced

In some islands of Polynesia, such as Tahiti and the Marquesas, breadfruit is the primary food source. This is more due to geography than anything else. On the steep mountain slopes of the more-eroded Marquesan islands, breadfruit could be cultivated, while taro could not. In Hawai'i, the gentle, fertile slopes of the younger islands were perfect for growing taro, and breadfruit was therefore a less important crop. Still, there were huge lowland groves of *'ulu* trees in Hawai'i. Among the first things mentioned by Archibald Menzies, the first Western botanist on Hawai'i Island, were the breadfruit plantations, which he stated were perfectly manicured. One fruit can weigh up to seven pounds and is high in starch as well as vitamins B and C. This crop was made famous by William Bligh's plot to bring breadfruit from Tahiti to the Caribbean to feed the slave trade. His mistimed adventure led to mutiny, and to the epic story captured in the book, *Mutiny on the Bounty*, which has been made into several movies.

WAUKE (vow-keh)
Paper Mulberry
Broussonetia papyrifera
Moraceae (Mulberry Family)
Polynesian-Introduced

Bark cloth production was an essential component of many different cultures at different periods in history, and the Hawaiians have been praised as possibly making the highest-quality bark cloth in the world, known as *kapa*. The plant they used was the paper mulberry, or *wauke*.

Making bark cloth was an arduous task that required a staggering amount of labor. Much of the labor went into the farming of the plant itself. As new stalks emerge from the trunks of the previous year's crop, the branch shoots and leaves had to be pinched off at their first appearance. The fibrous inner bark was used for making the *kapa*, and any branch left to grow would create disruptions in these fibers. A well-maintained *wauke* patch resembled a grove of odd palm trees with clean, branchless trunks and a tuft of leaves at the very top. The bark could be peeled off easily, and some Hawaiians would strip the bark while the tree was still standing.

After the outer bark was scraped off and discarded, the inner bark was cured with saltwater soaks. Women then pounded the bark with specialized, grooved beaters that pulled apart the fibers of the bark, greatly increasing the width, pliability, and softness of the material. From a one-inch diameter tree it was possible to produce an 18-inch wide piece of bark cloth that was sturdy, yet soft. The bark cloth was dyed, stamped with ink patterns, and scented with various flower extracts to produce beautiful natural cloth.

WET FOREST ZONE

THE WET FOREST is the most well-known habitat in Hawai'i, and people come from around the world to see the lush rainforests and tranquil waterfalls. A wet forest is classified by a minimum of 100 inches of rain per year, and there are ample areas that meet that qualification. Wet forests in Hawai'i can span from near the coast to over 6,000 feet elevation, where the moisture level drops off rapidly. In this broad range there are thousands of species of plants, ferns, birds, and insects.

The wet forests also creates specialized habitats. Waterways and bogs can spot the landscape, and rapid elevation changes can result in cloud forests constantly bathed in fog. In these super wet areas the water leaches nutrients out of the soil very rapidly, and the constant cloud cover restricts the amount of light. This leads to plants that are often stunted, dwarfed, or even bonsai-sized. There are also some unusual species that have adapted to live only in these habitats.

The plants in the wet forest are typically fast-growing in order to compete for the light. Ferns and mosses occupy every available habitat and can be found growing in soil, on rocks, in water, and on trees. The ferns in Hawaiian wet forests range from 30-foot, trunked specimens to nearly microscopic individuals. The ferns, particularly tree ferns, play an important role in the wet forests. They act as giant sponges and soak up water during the rainy times, then keep the forest moist during the drier periods.

To the Hawaiians the wet forest, especially upper elevation forests, were considered sacred, powerful, and occasionally evil places. Above the areas normally traversed by people was what the Hawaiians called *wao akua*, or the realm of the gods. The deep rain forest was not a place where Hawaiians would go lightly, but only with a specific purpose and using the proper ceremony and respect did one undertake the journey. One reason that Hawaiians entered the forest was to collect the small, brightly colored forest birds known as the Hawaiian honeycreepers. Believed to have come from a single founding population of an American finch, the honeycreepers evolved into over 50 distinct endemic species, an impressive example of adaptive radiation like that of the Galapagos Finches. The Hawaiians caught these birds live to harvest the 3–5 molting feathers at the base of the legs. These feathers were used to make royal capes and helmets, which could require up to 800,000 feathers.

'AMA'U (ah-mah-oo)
Sadleria spp.
Blechnaceae
Endemic

These are smaller tree ferns that are
usually found in drier climates than the
larger *hāpu'u*, in open areas and lava flows.
The trunks can exceed 6 feet in length, but
rarely do so. The new fronds of the *'ama'u*
are often an attractive bright red, gradually
fading into the dark green color of the
older fronds. The flesh of the *ama'u* stalks
was used to treat boils and other external
sores. The pith of the trunk is an edible
starch that was eaten during times of
famine.

HĀPUʻU (hah-poo-oo)
Hawaiian Tree Fern
Cibotium spp.
Dicksoniaceae
Endemic

These large ferns can reach heights of over 20 feet. The loose, fibrous "bark" on their "trunks" serves as an important nursery habitat for germinating forest tree seedlings and will often host many smaller species of ferns and epiphytes. The soft hairs, or *pulu*, that cover the base of the fronds of some species were used as a bandage, soaked in medicine and then applied to wounds to stanch bleeding and reduce infection. *Hapuʻu* ferns are still common in Hawaiian forests and are a popular landscaping item.

HŌʻAWA (hoe-ah-vah)
Pittosporum hosmeri
Pittosporaceae (Pittosporum Family)
Endemic

These small trees are usually found in upper elevation leeward wet forests. It is sometimes called the Hawaiian magnolia due to the tan fur that often coats the underside of the light green leaves. The small white flowers appear in densely packed, semi-spherical clusters that grow directly off the branches. The hard fruits were a favorite food of the *ʻalalā*, the Hawaiian crow. Now extinct in the wild, the crows no longer spread the fruits of the *hōʻawa*, which is becoming increasingly rare.

'IE'IE (ee-eh-ee-eh)
Freycinetia arborea
Pandanaceae (Screwpine Family)
Indigenous

ʻIeʻie is a climbing vine with thick, woody stems and tough aerial roots. Its long, dark green, serrated leaves and fountain-like growth pattern are reminiscent of the *hala* tree. Its huge, bird-of-paradise type flowers and large, bright red fruit are equally impressive. The aerial roots, when dried, are of incredible strength and were used in old Hawaiʻi to weave fish traps, sluice gates, and the framework for war helmets.

'ILIAHI (ee-lee-ah-hee)
Sandalwood
Santalum paniculatum
Santalaceae (Sandalwood Family)
Endemic

Sandalwood was once a dominant understory species in *koa* and *'ōhi'a* forests, but today can only rarely be found in deep valleys and other un-logged areas. The tree is actually parasitic, and underground its roots tap into the roots of other trees to steal nutrients. *'Iliahi* can reach heights of 30 feet and has light green leaves, sometimes with a yellowish or bluish tinge. The tree produces small golden flowers and purple, olive-like fruits. A flowering tree can produce thousands of these tiny flowers, and a grove of sandalwood trees sends a wonderful scent wafting through the forest. These flowers were added to the water used in making *kapa* cloth to infuse a long-lasting fragrance to clothing.

The crushed leaves have a sweet fragrance as well, but it is the wood of the tree that is popularly used for incense and was Hawai'i's first true export. Less than 20 years after Western contact, dozens of merchant ships stopped in Hawai'i each year to buy food and water on their way to China to trade for spices, silks, and opium. The merchants convinced the Hawaiian monarchy that they could fetch a fine price for sandalwood in China, and Kamehameha III, desirous of Western wealth, ordered many commoners to abandon their crops and instead harvest *'iliahi* from the forest. Huge expanses of Hawaiian rainforest were clear-cut to harvest the trees, the effects of which are still impacting ecosystems today.

KOA (koh-ah)
Acacia koa
Fabaceae (Pea Family)
Endemic

Koa is the largest tree in the Hawaiian flora. In remote rain forests there are still trees with 12-foot diameters that grow to well over 100 feet tall, and there are stories of even larger trees. The tree is easily recognized by its crescent moon-shaped leaves, which are actually not leaves at all, but flattened leaf stalks known as phyllodes. If you find a young *koa* tree you will notice that it has two different types of leaves, the crescent leaves and small, pinnate leaves, which are the true leaves. As the tree gets older it abandons the true leaves altogether and relies on its phyllodes for photosynthesis, a drought adaptation that *koa* is believed to have developed during an Ice Age period. As a nitrogen fixer, *koa* plays an important role as a nutrient provider in the forest.

The hardwood of *koa* is good for building, although the acidic sap of the tree makes it a poor wood for making bowls and other eating implements. The Hawaiians used *koa* almost exclusively for building their large, open-ocean canoes, which could measure over 100 feet in length. The task of selecting the proper tree was left to a forest bird named the ‘elepaio, a small flycatcher. If the ‘elepaio pecked at the tree, it was a sign that the tree was unsuitable as

it probably contained insects and would quickly rot. If the bird signaled that the tree was suitable by visiting but not pecking at the tree, it was felled and cleaned of its branches. The trunk was left to cure in the forest, shaped into a rough canoe hull, and then the trunk, which weighed many tons, was hauled by hand with ropes to the shoreline, where it was finished into a canoe.

Today koa wood is so valuable that very few are able to afford the hefty price tag that a *koa* canoe would represent. High-grade *koa* wood can sell for over $100 a board-foot on the commercial market. The size and age of a tree needed for a racing canoe, which is about 60 feet long, could equate to an overwhelming $500,000 in usable wood.

KOKIʻO KEʻOKEʻO
(koh-kee-oh keh-oh-keh-oh)
Hibiscus waimeae and *Hibiscus arnottianus*
Malvaceae (Mallow Family)
Endemic

These large shrubs can rarely be found growing in damp forest areas on Oʻahu, Molokaʻi and Kauaʻi. They both boast large and perfect white flowers virtually every day of the year. In addition, they have the distinction of being the only fragrant *Hibiscus* species on the planet. The species are distiguishable in that *H. waimeae* has more vigorous growth and overlapping flower petals, while *H. arnottianus* has a more compact growth and flower petals that do not overlap. One subspecies of *H. waimeae* is endangered.

KOKI'O 'ULA (koh-kee-oh oo-lah)
Hibiscus kokio
Malvaceae (Mallow Family)
Endemic
Rare

These small shrubs are rare in lowland wet forests on the larger islands, and most often Kaua'i. The flowers can occasionally be orange or even yellowish, but are primarily a beautiful bright red color. Like many of the *Hibiscus* species in Hawai'i, this plant could be used for fibers from the bark or in different medicinal mixtures.

KUPUKUPU (coo-poo-coo-poo)
Common Sword Fern
Nephrolepis exaltata subsp. *hawaiiensis*
Nephrolepidaceae
Endemic

Like most tropical climates, Hawaiʻi boasts a wide variety of ferns. Because fern spores are readily dispersed around the world via upper atmospheric jet streams, many ferns are indigenous, not endemic. However, because of its extreme isolation, 74% of Hawaiʻi's native ferns are endemic. The small sword ferns found in most wet habitats are an endemic subspecies of the more common and hearty common sword fern. These short, hairless ferns known as *kupukupu* are slowly being displaced by *Nephrolepis brownii*, an invasive fern that is much hairier with frond segments that are longer and brittle.

LOBELIADS
Brighamia spp., *Clermontia* spp.,
Cyanea spp., *Delissea* spp.,
Lobelia spp., *Trematolobelia* spp.
Campanulaceae (Bellflower Family)
Endemic, Endangered

The lobeliads have been described by
early botanists as "the pride of the
Hawaiian flora." The term *lobeliad* refers
to a loose collection of plants within
the Bellflower family, and are represented by six genera, five of which are endemic,
comprising 108 species, all of which are endemic. Beautiful bellflowers are displayed
one at a time by some species, while others boast hundreds of flowers at once.

These remarkably diverse plants have co-evolved with many of the native nectar-feeding
birds that have beaks specifically adapted to fit each species' flowers. This unfortunately
means that the decline and extinction of both plants and pollinating birds may go hand
in hand…if one is lost, so is the other.

Lobeliads are generally short trees or shrubs that can grow either epiphytically or
terrestrially. They are usually unbranched or sparingly branched and the leaves tend to
cluster at the tips of the branches, lending the plants a palm-like appearance. The flowers
all have petals that fuse into a tubular formation, giving the Bellflower family its name.

MAILE (mai-lay)
Alyxia stellata
Apocynaceae (Dogbane Family)
Indigenous

Maile can grow as a twining shrub up to several feet high or as a true vine. Its leaves are smooth, leathery, and shiny. *Maile* is a favorite in Hawaiian lei-making due to its sweet smell, which is produced by coumarin found in the leaves and stems. A typical *maile* lei is made by removing the inner stem of three strands of vine and twirling them together. Because of *maile*'s extensive use as a lei it was one of the five plants sacred to the hula altar and was often used in hula, both as adornment and as an offering.

MĀMAKI (mah-mah-key)
Hawaiian Nettle
Pipturus albidus
Urticaceae (Nettle Family)
Endemic

Māmaki is a small tree common in the understory of mesic and wet forests of the major Hawaiian Islands. The plant has papery, ovate leaves with reddish veins and stems, and white globular fruit growing directly on the branches. These small fruits are edible and were among the few edible fruits on the island at the time the Hawaiians arrived. A natural tea can be brewed from the leaves, and is still used for relaxation and vitality when fully dried, and for muscle cramps and menstruation problems when green. In addition to being medicinal, it is a fine-tasting tea—much like Chinese green tea. The bark of the *māmaki* is used to make both cordage and *kapa*.

This plant is also larval host to one of only two native butterflies in Hawaiʻi. Known as the Kamehameha Butterfly *(Vanessa tameamea),* this monarch look-alike can still occasionally be seen around Hawaiian forests. The caterpillar lives on the *māmaki* plant, where it cuts and folds a leaf to form a small "house." The larger catepillars are impressive, inch-long, spiny green larvae. The adults feed on a variety of plants and nectars, but are said to prefer the sweet fresh sap of the *koa* tree.

MĀNELE (mah-neh-leh) or **A'E** (ah-eh)
Soapberry
Sapindus saponaria
Sapindaceae (Soapberry Family)
Indigenous

The *mānele* is a large tree that can commonly reach 80 feet in height. The divided leaves are reminiscent of various ash species, although the soft wood of the *mānele* was rarely used for timber. The common name, soapberry, is derived from the sudsy water that can be made from soaking the ripe berries in water. The hard, round seeds found inside each berry was highly prized for lei-making and other decorations. When polished and oiled the shiny beads are a beautiful glossy black.

'ŌHI'A 'AI (oh-hee-ah eye)
Mountain Apple
Syzygium malaccense
Myrtaceae (Myrtle Family)
Polynesian-Introduced

This tree is commonly called
mountain apple because of the
sweet, apple-like fruit that forms
on its branches. The fruit, although
delicious, has a very short shelf life
and has to be eaten within hours
of picking them fresh off the tree.
There are two varieties of *'ōhi'a 'ai*,
one with white flowers and white
fruits and another with bright pink
flowers and red fruits. This tree is
commonly found in the understory
of wet forests and is a favorite of

local Hawaiians, who pick the fruits with long bamboo picking poles. The mountain apple tree also produces a dense wood, relatively light in color, which was commonly used for carving large wooden images, known as *ki'i*. The bark of the plant was also used medicinally for external infections and throat disorders, including bronchitis.

OLONĀ (oh-loh-nah)
Touchardia latifolia
Urticaceae (Nettle Family)
Endemic

This shrub was a common addition to the understory of Hawaiian wet forests, but has become increasingly hard to find. *Olonā* was prized for the high quality fibers that could be harvested from the inner bark, which were used to make a variety of cordage for fishing and other activities. *Olonā* fiber, under more scientific scrutiny, has been found to be one of the strongest plant fibers in the world, with its combination of tensile strength and fiber length.

PĀPALA KĒPAU (pah-pah-la keh-pow)
Pisonia brunoniana
Nyctaginaceae (Four-o'clock Family)
Indigenous

These small trees can be found in dry and moist forest areas of the larger islands, except Kauaʻi. The small white flowers are borne on stalks and precede the formation of elongated fruit. The fruit, which are present during autumn, are covered with an extremely sticky substance, akin to partially dried rubber cement. This sticky coating was harvested and smeared onto tree branches directly beneath a flower that was food for a desired bird. The bird would eventually land on the branch to drink from the flower and become stuck in the process. The Hawaiians could then harvest the feathers and release the bird, hopefully to catch it again during the next molting season.

ʻUKIʻUKI (oo-kee-oo-kee)
Dianella sandwicensis
Liliaceae (Lily Family)
Indigenous

These low-growing members of the Lily family can be found in a broad range of wet habitats on all of the larger islands. The long, slender leaves are tough and flexible and were traditionally used for making cordage. Hale Pili, the grasshouse

at Bishop Museum is lashed with ʻukiʻuki cordage. The attractive berries appear on stalks and make dyes that vary from light blue to violet.

AMY GREENWELL AND
HER GARDEN LEGACY

Tucked away in Kealakekua, Hawai'i is the Amy B. H. Greenwell Ethnobotanical Garden, a living monument to one of the outstanding women of the 20th century here in Hawai'i. Though the Garden commemorates her name and her property, her life history and accomplishments are rarely commented on, despite their rather astounding nature.

Amy Beatrice Holdsworth Greenwell was one of the 23 grandchildren of Henry Nicholas Greenwell, a soldier-turned-merchant who arrived in Hawai'i in the 1850s. Despite a series of rough starts, Henry found fortune in South Kona, where he ran cattle and a general store. Most famously, he is credited with giving Kona coffee its renowned name by winning a recognition award for his Kona coffee at the 1873 World's Fair in Vienna, Austria. He was able to bestow upon his family a 36,000-acre legacy when he died, a portion of which eventually passed on to Amy.

An observant and intelligent student, Amy possessed a passion for Hawaiian studies. She interrupted her education at Stanford University in 1942 to serve as a Red Cross nurse at Queen's Hospital during WWII. After the war she traveled to New York, where she worked with Otto Degener at the New York Botanical Garden on one of the authoritative volumes on Hawaiian plants, *Flora Hawaiiensis*.

Amy Greenwell had a great appreciation for both the natural environment and the Hawaiian culture, and on her return to Hawai'i in 1947, she started working closely with Bishop Museum and its archaeological projects. On one of her expeditions in 1953, she discovered some ancient fishhooks at Ka Lae and in doing so brought the now-famous Pu'uali'i sand dune site to the attention of professional archaeologists. This site led to the discovery of some 1,600 fishhooks of 65 varieties that were carbon dated to 950 A.D., which at the time was the earliest dated archaeological site in Hawai'i. In addition she did archaeological and botanical surveys for such significant sites as Pu'uhonua 'o Hōnaunau and Lapakahi.

Despite her archaeological accomplishments, Amy Greenwell was best known as a botanist. Over her lifetime, she wrote many articles on both native and other tropical plants, some of which include *Taro—With Special Reference to its Culture and Uses in Hawaii, Rose Growing in Hawaii,* and *Hawaiian Violets.* In her later life she lived at her 10-acre Kealakekua property. She began to transform this land into a "pre-Cookian" garden, her own term implying the days before Captain Cook, the first recorded white man in Hawai'i. She planted scores of native and Polynesian-introduced plants among intact remnants of Hawaiian agricultural formations.

In her free time, Amy dabbled in other sciences such as meteorology, the study of weather patterns, and speleology, the exploration and study of cave systems. On these subjects she wrote various articles and even gave regular weekly radio broadcasts. She reported surface weather and UAP, or unusual aerial phenomena, to the National Weather Service for over twenty years. In addition to all this, she raised pure-bred pug dogs.

When she died in 1974 at the age of 53, she left her Kealakekua property to Bishop Museum as an educational and cultural resource for locals and visitors alike to revisit the Hawaiian past and see the environmental splendors that ancient Hawai'i had to offer. Bishop Museum has expanded on Amy's initial efforts to encompass over 200 native plant species, many of which are on the endangered species list and a couple of which are virtually extinct in the wild. The now 15-acre garden is open to the public weekdays between 8 a.m. and 5 p.m. Guided tours are scheduled at 1 p.m. every Wednesday and Friday for $5 and a free tour is offered the second Saturday of each month at 10 a.m. Thematic tours are available for groups of five or more focusing on topics such as the Hawaiian agricultural field systems, Hawaiian herbal medicine, Polynesian migration, and the natural history of Hawai'i. Please contact the Garden at (808) 323-3318 or agg@bishopmuseum.org to reserve your tour.

INDEX

'āweoweo, 52

Bacopa monnieri. See 'ae'ae

banana. *See* mai'a

beach morning glory. *See* pōhuehue

beach pea. *See* nanea

beach vitex. *See* pōhinahina

Bidens hawaiensis, 65

Bidens menziesii, 65

Bidens micrantha, 65

Bidens spp. *See* ko'oko'olau

Blackburn Sphinx Moth, 48

breadfruit. *See* 'ulu

Broussonetia papyrifera. See wauke

Caesalpinia kavaiensis. See uhiuhi

Calophyllum inophyllum. See kamani

candlenut tree. *See* kukui

Capparis sandwichiana. See maiapilo

Chamaesyce celastroides. See 'akoko

Chamaesyce olowaluana. See 'akoko

Chenopodium oahuense. See 'āweoweo

Cibotium spp. *See* hāpu'u

Clermontia spp. *See* lobeliads

coastal zone, 7

coconut. *See* niu

Cocos nucifera. See niu

Colocasia esculenta. See kalo

common sword fern. *See* kupukupu

Cordia subcordata. See kou

Cordyline fruticosa. See kī

Curcuma longa. See 'ōlena

Cuscuta sandwichiana. See kauna'oa

Cyanea spp. *See* lobeliads

Cyperus javanicus. See 'ahu'awa

Cyperus laevigatus. See makaloa

Delissea spp. *See* lobeliads

Dianella sandwicensis. See ʻukiʻuki

Diospyros sandwicensis. See lama

Dodonaea viscosa. See ʻaʻaliʻi

dry forest zone, 45

ʻelepaio, 108

endangered, 34, 45, 48, 53, 54, 62–63, 66, 70–71, 72, 76, 110, 113, 127

endemic, 10, 11, 19, 22, 23, 25, 29, 34, 35, 36, 41, 49, 52, 53, 54, 56, 58, 62, 65, 66, 67, 68, 70, 72, 73, 74, 76, 99, 100, 101, 102, 106, 108, 110, 111, 112, 113, 117, 122

    definition, 4–5

Freycinetia arborea. See ʻieʻie

Gossypium tomentosum. See maʻo

hala, 15–16, 104

hala pepe, 53

hāpuʻu, 100, 101

hau, 17

hau kuahiwi, 54

Hawaiian caper bush. *See* maiapilo

Hawaiian cotton. *See* maʻo

Hawaiian crow, 102

Hawaiian Islands,

    arrival of Polynesians to, 5

    formation of, 3

    natural history, 4–5

Hawaiian nettle. *See* māmaki

Hawaiian poppy. *See* pua kala

Hawaiian tree fern. *See* hāpuʻu

Heliotropium anomalum. See hinahina

Heteropogon contortus. See pili

Hibiscadelphus hualalaiensis. See hau kuahiwi

Hibiscus arnottianus. See kokiʻo keʻokeʻo

Hibiscus brackenridgei. See maʻo hau hele

sweet potato. *See* ʻuala

Syzygium malaccense. See ʻōhia ʻai

Tacca leontopetaloides. See pia

taro. *See* kalo

Tephrosia purpurea. See ʻauhuhu

Thespesia populnea. See milo

ti leaf. *See* kī

Touchardia latifolia. See olonā

Trematolobelia spp. *See* lobeliads

turmeric. *See* ʻōlena

ʻuala, 79, 94

ʻuhaloa, 75

uhiuhi, 76–77

ʻukiʻuki, 125

ʻūlei, 42–43

ʻulu, 95

Vanessa tameamea, 117

Vigna marina. See nanea

Vitex rotundifolia. See pōhinahina

Waltheria indica. See ʻuhaloa

wauke, 96–97

wet forest zone, 99

Wikstroemia uva-ursi. See ʻākia

Zingiber zerumbet. See ʻawapuhi